Amiable Little Beasts

Investigating the Lives of Young Animals

Conceived and Edited by Ann Guilfoyle

Text by
Roger A. Caras and Steve Graham

MACMILLAN PUBLISHING CO., INC., New York

An AG Edition

All animals are at some time young. The systems that bring them into the world and which allow them to grow up are what **Amiable Little Beasts** *is about. The youngsters shown in the first pages (in order of appearance): white-tailed fawn, school of baby tropical fish in their coral environment, leopard kitten, infant chimpanzee, herring gull chicks, mother raccoon and baby in den tree, bobcat kitten, prairie dog pup.*

The authors wish to dedicate their text to Pamela

Macmillan Publishing Co., Inc.
866 Third Avenue, New York, N.Y. 10022
Collier Macmillan Canada, Ltd.

10 9 8 7 6 5 4 3 2 1

Loren Eiseley quote from *The Night Country*
(New York: Charles Scribner's Sons, 1971).

Library of Congress Cataloging in Publication Data
Main entry under title:

Amiable little beasts.

1. Animals, Infancy of. 2. Parental behavior in animals.
I. Guilfoyle, Ann.
QL763.A44   591.3'9   80-15367
ISBN 0-02-546560-0

Printed in the United States of America

. . . they are delicate and not
in a position to defend
themselves. So I look out
for them. I'd do as
much for you.

— Loren Eiseley

Contents

Introduction

When we speak or write about nature, there is the tendency to oversimplify, to assign true purpose to what may well be called random happenings. Yet, in the animal kingdom, one generalization does seem to hold true, and one truth overrides all others: apparently the first, the paramount, the absolute rock-bottom fact of life is that, in all its forms, life strives to perpetuate itself. Each animal's biological purpose for coming into being is to reproduce itself in kind. And each creature is a carrier of the stuff of its kind, a biological envelope fashioned to sustain heritability—at the right temperature and with an appropriate supply of the right nutrients—until called upon to briefly interact with other matched substances to create new life. But reproduction is just a beginning, meaningless unless the new young are able to develop into functioning adults ready, in turn, to pass along the unique characteristics of their species to yet another generation. The ways that have evolved to achieve this perilous, all-important passage from youth to adulthood are enormously complicated, taking many hundreds of thousands of different forms. And that is what this book is about—the systems that support and sustain young life in which all future life is waiting.

To some degree the world is hostile to all new life. That is true for many reasons. Life outside of the egg is shocking when compared to the snug, moist warmth within. A young animal facing the outside world, whatever its method of birth or hatching, no matter where it occurs, faces variability for the first time. Life in the uterus or in the egg case is always the same. Life outside constantly changes. An animal's ability to adapt to the changes as they occur, and to survive them, is the first test of that animal's suitability to carry forward the inherited characteristics of its kind.

Those early tests to which an organism is subjected could be *too* harsh, *too* demanding; but if that were so, too few individuals would survive to give a species very much of a chance. Many animals, particularly at the upper end of the animal kingdom, have developed enormously elaborate social structures: devices and systems to allow parents, siblings, or surrogates from the social group to care for the young and helpless. The largest of mammals—whales in the sea, elephants on land—would be lost within hours of birth if not supported and protected by more mature specimens of their own kind. The gorilla would not survive its first day on earth if not meticulously cared for. Few animals are as helpless as the great predators when they are born. Tigers, lions, leopards, bears—all of them would die within hours of birth unless there was an elaborate plan, peculiar to each species, to help them make it through those early hazardous months.

Sometimes closely related animals have evolved different schedules for the same events, but the events remain in parallel, for the outcome must be the same: the young growing and becoming independent. Rabbits and hares are both lagomorphs, yet rabbits are born hairless and helpless, unable to accept any food except their mother's milk, while hares are born with hair, with their eyes open, able to hop about and nibble vegetable matter very shortly after birth. Rats and mice are pink and helpless at birth; guinea pigs, also rodents, are quite precocious and take solid food almost from the start.

In the pages that follow, using the work of thirty-two of the finest wildlife photographers and text that is a synthesis of what is known (or at least believed) about the social systems that support the animal species that are left, we explore the world of the survivors, those young born equipped to face all the obstacles nature puts before them. There is no doubt that the very obstacles are selecting devices. Only the salmon able to leap the falls can reach the gravel bed that is the only place where it can reproduce. Every young animal growing up faces innumerable waterfalls, rushing streams, seemingly insurmountable barriers that stand between it and the moment when it will fulfill its goal and help to produce the basis for the next generation.

The greater the early-life hazards and the less likelihood there is of an animal reaching maturity so that it can reproduce, the likelier it is that more young of its kind will be produced. A codfish may lay three million eggs, while an elephant will bear a single young after nearly two years of gestation. When the degree of hazard increases so rapidly that a species cannot compensate for it with an increased reproduction rate, the survival of that species is unlikely. In the last two hundred years the world of the California condor has become far more dangerous than it once was. Yet the condor, a creature of instinct, has no choice; a few mated pairs in a small surviving population still produce only one egg every second year. That reproduction rate suited a time without off-road vehicles, without firearms, without well-cared-for cattle dying too slowly in open-range conditions; it was a time without pollution, oil prospecting, or poisons. The fact that the condor has a reproductive rate for a world that no longer exists and can never exist again has brought the species to the edge of extinction. Condors must now be taken and bred in captivity. No animal whose reproductive rate and style belong to a lost world can survive for long on its own. The condor is just one of many threatened by change.

From the very start, it is an upstream race of olympian proportions. A single male bison, for instance, may produce between five and ten billion sperm every time it mates. Only one of those sperm need make it to the uterus, there to find an egg. Only one sperm need fertilize the egg so that it will implant itself in the uterus, there to grow and become, one day, a bison too. The odds are not always one in billions, but they are always great enough so that the nurturing, caring for, and training of the young is the great task of animal societies, be they no larger a group than mother and young, or as large as a school of herring. For animals without families or without a social grouping that can be instructive in social skills, greater reserves of instinctive behavior must be provided. All animals, including man, have instincts, but in the higher animals the amount of dependence on innate behavior decreases. For the worm, instinct will be nearly 100 percent of its guide through life. It is very far below that mark for chimpanzees, lions, and wolves.

Whether there is intense family care or no family care at all, for all animals there is a constant. Enough young in each generation must be provided for and be complete enough packages in themselves to resist the forces that try to destroy them. Predators, weather, disease, misadventure of all kinds, lie waiting. Somehow, from each species, enough young must pick their way through the traps and pitfalls to reach maturity and begin the cycle anew.

Roger A. Caras
Steve Graham

New Life

It is ironic that one of the most ordinary of all occurrences touches us most deeply. We are more sentimental about birth than we are about any other biological happening, even death, yet one is just an extension of the other. Birth, like Spring, means beginning, and we humans look on beginnings as celebrations, as reaffirmations, as proof that things can go well with the world. We see in almost all newborn or newly hatched animals a reflection of our own being, our own hopes and joys when we, too, extend ourselves forward into time by having young. If there is anything in nature about which we cannot be objective it is this, the starting over and reaching ahead signaled by the appearance of a new creature to pick up the challenge of species survival. Baby animals attract our affection as no older animals can, and mothers performing their roles, in species where that is part of the scheme, earn our admiration.

Perhaps our lack of objectivity about animal young is a good thing, if it evokes the universal reverence for life called for in so many of our philosophies but so seldom encountered. Emotional reaction may be poor science, but it can be an important aid to furthering the survival of the animals that science would like to preserve and study.

Tadpoles, each in their own life-support capsule, and salmon (overleaf) in their egg-cases await the shock of hatching. The survival rate of the young will not be high, but it will be high enough in each case so that their species will get through another generation intact.

A herring gull chick and baby hognosed snake break through the protective shell of the egg. All birds and some snakes hatch from eggs. (A very few primitive mammals do, as well.) It is strange that birds, which are descended from reptiles, should all be tied to the more primitive egg, while some reptiles have become viviparous (live-bearing). Species still using the egg are oviparous.

A box turtle hatches. A complete and perfect animal in miniature comes forth to assume its role in the cycle of birth, growth, and reproduction. Because it does not belong to a society-building species, the turtle must be born with all of its survival information intact and available for instant use.

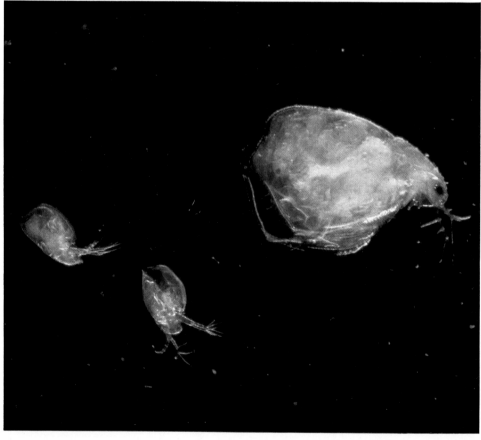

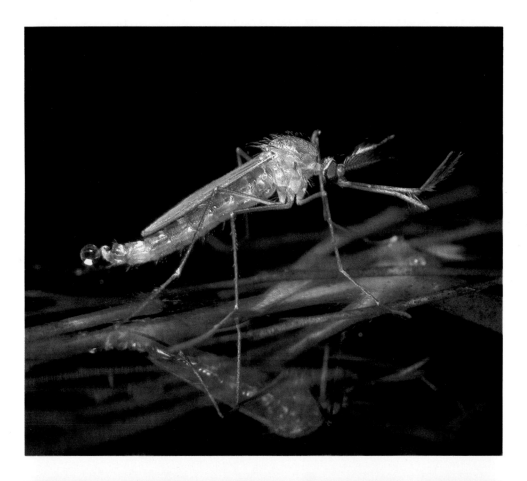

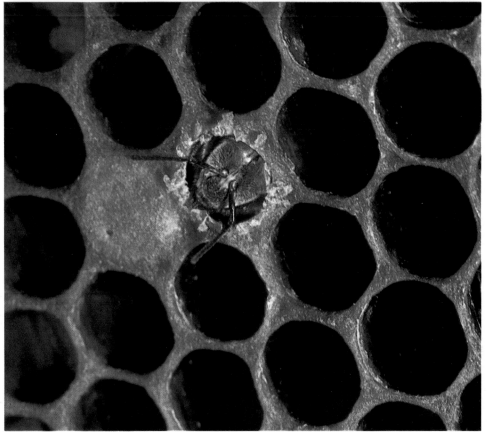

Styles of birth: an aphid (left above) and waterflea (left below) give birth to living young, a mosquito emerges from its pupa, a honeybee moves out of its nursery cell into the hive, and (overleaf) a viviparous garter snake with her large brood. Different answers to the same problem—how to keep everything that has gone into a species moving forward generation after generation, in relentless pursuit of eternity. The differences we perceive are often more in style than in substance.

A white-tailed deer immediately after birth. The very act of giving birth will trigger patterns of instinct in the doe. The fawn is born with instinctive reactions to those patterns. The doe will know how to signal her fragile infant to remain quiet and still and wait for her return. The fawn will know to obey and how to obey. A failure on either's part will spell disaster for the fawn.

Animals that are preyed upon by many are especially fruitful in the production of offspring. Deer mice (overleaf) are prolific and quick to mature because almost every meat-eating animal in their vicinity will hunt them as a staple in its diet.

An animal's role in life is not anticipated by its nature in infancy. Indeed, in the case of the prairie falcon future power is belied by the infant state. Newly hatched prairie falcons are as vulnerable as the young of their future prey.

First Days

Far, far more young are born to most species than need grow up, can grow up, or would, in fact, be useful as adults. Perhaps 50 percent of all lion cubs fail to reach maturity, showing that this is not limited to prey animals but includes the great predators themselves. When we see a single impala buck with his harem of thirty or more females, we are reminded of the surplus of males that were produced. The surplus males did not survive, but fed the predators that always appear where impala are found.

Young rabbits huddling in their nest will almost all die long before they are ready to breed. Snakes, hawks, eagles, owls, wild canids and felids all take their toll, and now that there is man, vehicles and feral cats and dogs join in the slaughter. Only a small percentage is supposed to survive. That is the design.

The terrible vulnerability of young animals makes such perfect sense that we must admire the system even while we despise it. Those young that grow toward maturity are, almost always, far better examples of their kind than those that perish. They *deserve* to survive and their superior qualities merit reproduction. This is not always true, of course, for there is chance, the thing we call bad luck. But it is true often enough to constitute an elemental rule of survival for every species. The best must make it, the least must fall by the side of the road, for only in that way can a species remain strong and viable. A niche in nature is not easy to hold. Only the best can do it.

Virtually everything a young animal is, does, and seems to be has as its purpose some degree of compensation for its extreme vulnerability during infancy. The color, shape, and posture of a young fish camouflaged against a background of coral are designed to protect it from predators. Newly hatched robins (overleaf), although blind and completely helpless, instinctively know how to beg, how to stimulate parental concern and launch hunting forays that will mean food brought back to the nest and their own survival.

*Infant mammals (   )
most helpless of (   )
baby opossum (le   )
weeks old is able (   )
hanging onto its mo(   )
it fails to clin(   )(   )
enough it could be dropped and
abandoned. The raccoon is ten
days old, its eyes and ears are
still sealed shut. As an adult the
raccoon is a survivor; it seems to
be able to live anywhere. As an
infant, though, it is a very fragile mote of life.*

Very often we see in the faces of young animals their absolute dependence on parents and fate. More and more they depend, as well, on us. The baby harp seal is so appealing that its face has launched a fierce protectionist battle and an international outcry against its slaughter. Baby short-eared owls, seemingly all eyes, belie the powerful night hunters they will grow to be.

Those species for whom no parental care is provided know how to set off immediately on their own. They must do or die as soon as they are born or hatched. But those species designed to be cared for know to wait and expect and respond. Deer mice (left above), cottontail rabbits, and young Canada geese seek warmth and contact from each other, physical and emotional necessities, while waiting for the return of their parents.

43

Cryptic behavior is found throughout the animal kingdom, and no animals are better served by the high art of hiding than the young. To protect itself while its mother is away the infant caribou instinctively curls up on material most like itself in color and texture, in this case lichen-covered rocks. It will, if it lies perfectly still, be almost impossible to see. The mountain lion (overleaf) is spotted, for it will be left against a dappled background when its mother must go off to hunt. As it matures the spots, no longer needed, will fade.

45

Parental Care

In the amount of care young animals receive, nature has run the gamut. Sea turtles hatching in the sun-warmed sand of a traditional nesting beach, salmon working their way free into gravel-bottomed streams, never know a parent. To them a father and a mother are biological necessities, only that.

Then there is the other extreme, an infant gorilla cradled in arms as solicitous as human arms greeting a newborn baby. The female elephant and, in fact, the entire herd show endless concern for the young. A bobcat mother will accept any challenge in defense of her young.

There are systems in between, of course, but interestingly enough, most species tend toward either extreme. It is generally in the direction of all or nothing at all.

Where parental care does exist it is, for the human observer, one of the most enchanting aspects of natural history. To us it is right for a mother to care for her babies. That is the way we are, ideally, and therefore we see it as a reflection of ourselves at our best. However, it is not this simple. Where parental care exists, it is because it has evolved as the best way for the species to survive. But this is equally true when the youngster is totally independent and neither knows nor needs parental care. If it works it is right. If it doesn't, then extinction is not very far away.

Until her eggs were ready to hatch, the wolf spider dragged behind her the small sac that contained them and guarded it with fierce determination. Then she ripped the sac open, freeing the young that clambered onto her back for safety. They will remain clinging to their mother until they are able to cope with the predators that can make even a young spider's life one of great hazard.

Although the newborn Grant's zebra is able to stand very shortly after birth, barely stopping long enough to catch its breath before starting to struggle to its feet, it is still helpless. It must have its mother's milk and learn from her and the rest of the herd what is dangerous and what can be ignored. The jackal in the foreground is not a threat, but other predators are. The zebra will learn one predator from another by watching the reactions of the older animals.

The Kodiak bear is the largest meat-eating land animal left on earth, yet Kodiak cubs are vulnerable and protected only by a fiercely determined mother. No less devoted is this long-billed curlew hen. Young birds have trouble regulating their body temperature during their first days out of the egg. The curlew mother not only feeds, hides, and eventually educates, but also protects her chicks with her body against sudden temperature changes that they could not handle themselves.

It is very difficult to understand why or how nature divided parental duties. In some species it is the female alone who sees to the young, in fewer species the male carries the full load, and in still others both sexes participate. With flickers and pelicans both parents pitch in. A male flicker (left) enters the nest cavity to bring food to his always demanding young. A female pelican regurgitates food to her greedy chick who thrusts not just its bill but its entire head into its mother's cavernous mouth.

55

Within an hour of birth a bison calf tries to feed. It does not instinctively know where its mother's teats are but must find them. Very often the search will take the calf to the area between the front legs with predictably disappointing results. In time, though, the calf learns where the milk is to be had and, having successfully absorbed one of life's first lessons, will be linked to its mother for as long as she will tolerate its demands.

Mammals that live under water are able to nurse there without difficulty. Obviously, when land mammals moved back to their natal sea this adaptation was required. If the manatee seems less a mammal because of its shape and habitat one need only see an infant suckling to be reminded of the true nature of the species. Manatee milk is exceptionally rich in fats, facilitating rapid growth in the young.

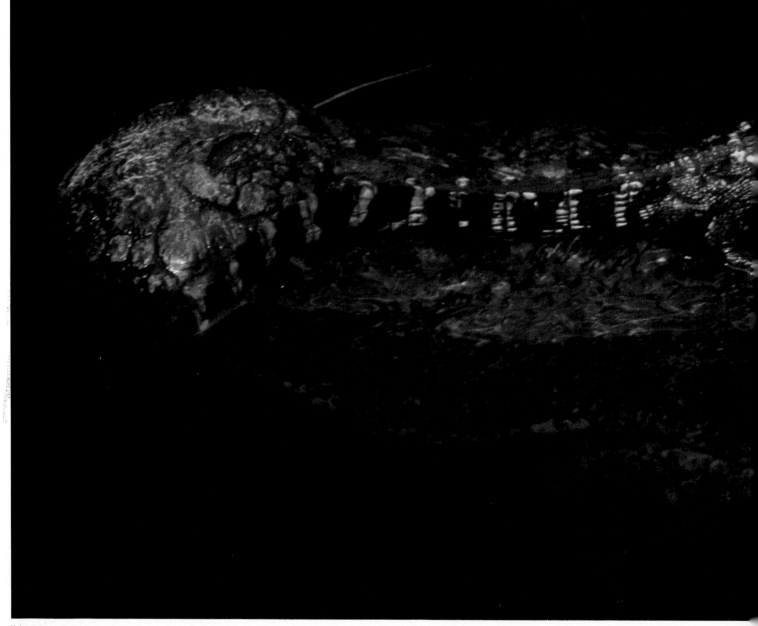

There has been some controversy about parental care among crocodilians and, indeed, many observers were adamant that there was none at all. In fact, it was claimed that mother alligators ate their own offspring if they caught them. It is difficult to understand how such misconceptions arose. Mother alligators protect their eggs and do provide haven for their young. Here a baby rests on a female's snout.

Close physical contact is a hallmark of maternal care. Young brown bats sleep close to their mother, and deer mice are moved to safety in the mouth of an ever-solicitous mother. The lioness (overleaf) uses the same technique as the deer mouse to transport her infant. One of the first lessons any young animal subject to parental care must learn is not to resist. In nature, whatever the parent wants must be accepted without serious argument.

No infant and mother combination comes as close to home as the primate pair. We hear so often that "they look almost human." In fact, the care of babies by primate mothers is rather like our own except that the monkey is likely to put up with less nonsense from her babies. Monkey mothers can be serious disciplinarians.

Canada geese mate for life, or for as long as both survive. They form a tight family unit when the goslings hatch, and any predator intent on molesting these babies will have to reckon with double parental wrath. A hissing, charging goose in defense of his or her own young is a formidable opponent.

The lifestyles for which the baby eagle and the baby harbor seal are being readied could hardly be more different, but they have in common both the attentiveness of at least one parent and their absolute dependence on it. Neither eagle nor seal could survive for much more than a day without the food provided them. The seal's mother will provide her own milk, the eagle chick will be fed food gathered in the hunt.

All young mammals seem to find their parents sources of never-ending wonder. The babies never tire of looking, sniffing, and nuzzling the larger versions of themselves. The scent, sight, and sound of a parent are critical pieces of information for the young to have. Once these recognition points are established they will function as bonds to keep the young close to the adult. The baby silver-backed jackal and the young mountain goat are temporary satellites of their parents, bonded to them until they are able to survive on their own.

Grooming, licking, cleaning are constant activities among mammals. It is less hygiene than bonding that is involved. Youngsters must know their mother instantly and must focus on her no matter what else offers itself as a distraction. We cannot clearly define emotions in animals, and although we are tempted to say the mother mule deer loves her baby, there is a practical consideration in the attention being given. It is reinforcement of a bond essential to the fawn's survival.

Both the elephant and the black rhino must protect their young, for they live in an area of powerful predators. Although as adults they will be immune from attack, as young they are susceptible. Trouble can occur only if the young wander too far away from their mothers, something neither is likely to let happen. Any area left in Africa remote enough to support elephant and rhino almost certainly will support lion as well. It is a threat that never goes away.

Although groups of giraffe tend to spread out when they feed, the young are never very far from their mothers. The reticulated giraffe seen here may seem otherwise involved but the safety of her calf is her first concern. If she should start at the sight of another animal and move off, the baby dogging her steps will move off right behind her and will learn what it was that frightened her. The experience will be remembered and, for as long as it lives, sight of the threatening animal will be cause for alarm.

Growing Up

For those species that will develop under parental guidance the growing period is one of learning. For those species that are on their own from the beginning the amount of learning involved is negligible. A newborn rattlesnake knows without hesitation to strike when threatened. What's more, its venom is, drop for drop, more toxic than that of an adult, compensation for the smaller amount produced by infant-sized glands. A young insect or spider can learn to avoid a hazardous course, but little more than that. Where parental guidance and herd example are not part of the plan the survival package must be virtually complete as the young animal emerges.

Animals born to learn do so with a variety of teachers. More than personal experience is involved; there is parental and often herd experience to be gleaned as well. A young chimpanzee fears a forest cobra because it watches the adults of its troop react with fear. A baby raccoon watching its mother overturn rocks to find crayfish will soon imitate her and find its own food. A young timber wolf learns quickly to defer to the dominant male, a social grace that is essential to the smooth operation of the pack.

Even the play of higher animals is an essential element in their growing up. A lion cub playfully stalking a littermate is sharpening the skills that will later be used in the hunt. Adolescent bisons testing their new horns are practicing for the later contests that will win them the right to mate. As the young animals play and grow together the rules are safely learned and the groundwork for adult survival is laid.

A snowy owl, partially grown but still with infant down, has reached the awkward stage. Its demands on its parents are staggering. It is always hungry because it is always growing. By the time the young owl is ready to go off and feed on its own its parents will be near exhaustion.

The young of long-legged animals, like this zebra, always look ungainly. That is because they are. The long, seemingly delicate running legs that mean survival on the African savannah will not be in proportion to the animal's body or be very well coordinated until it is nearly grown.

As it grows, the white-tailed fawn enters a period of particular danger. It cannot yet run fast enough or jump high enough to escape a determined predator. Spotted camouflage is its only protection as it begins to explore its world.

Animals that live high, like this bighorn sheep (right) and Rocky Mountain goat require special skills that must be learned at the very outset. Shortly after birth they will demonstrate incredible balance and fine coordination, but the young mountain climbers must still learn to handle the heights. A mistake in judgment could cost the animal its life.

A grizzly bear at three months has already done an extraordinary amount of growing, since it was little bigger than a mouse at birth. It is destined to be a giant of 800 pounds or more. That enormous growth must be well advanced before the second spring of the young grizzly's life when its mother will send it off on its own so that she can mate again.

The baby wood duck must reckon its growth to independence in days and weeks, not months and years. It is hatched in the spring and will be off in the fall. The correct timetable for development is established for each species, and all young animals must grow and develop on schedule or perish. There is very, very little room for deviation.

83

The Couch's toadlet (left) and the striped-neck musk turtle must grow and survive without benefit of enormous brain power. They do not have parental guidance, they are slow-witted, and their world is no less hazardous than the bird's or the mammal's. Still, enough appear every spring to allow for the inevitable attrition and many of them make it to maturity.

The opossum is a dull-witted animal at best, but it is tenacious, fertile, persistent as a species, and therefore very, very successful. While other mammals fall victim to man's manipulation of the environment, they thrive. A lot of young, dull-witted opossums grow up to perpetuate their kind. Brains aren't everything.

Those animals, such as the chimpanzee and raccoon, that are going to be to some extent arboreal have special lessons to learn and skills to develop. Trees grow high, and moving through them, especially when disturbed by predators, can be dangerous in itself. Balance, depth perception, judgment, coordination, and strength all must come early and continue to develop.

Prairie dog pups practice a ritual
necessary for animals that live
their lives in a prairie dog community—nose contact. Growth
for a social animal is not just a
matter of building bone and muscle but a time of mastering the
manners and responses of their
own society.

Play is a learning system and a survival device. The rough-and-tumble engagement of these two young raccoons sharpens senses and reflexes, reduces fear of physical contact, and prepares the young animals for when they may have to contest others of their kind for food, space, and mating opportunities.

For two red fox cubs the world outside the den is both tempting and threatening. Sounds, smells, and light have come in from the outside since their senses first awoke. Their desire to explore the world is part of the intelligence that will sustain them in the dangerous world all foxes face.

Young bears of the same litter stick together in one of the three social groupings they will ever know. The first is with their mother, then with each other, and later with their mates. All their associations are destined to be transitory.

The moose is the largest member of the deer family and as an adult may be a very dangerous adversary even for the wolf (the moose's only enemy besides man). But for the moose calf the world is an awesome place. There is a sense of safety in contact with a known quantity, like another young moose.

Even on the eve of their departure from the nest, young barn swallows (overleaf) frantically signal their parents for food. For a period ranging from 18 to 23 days both parents work constantly to meet the demands of their brood. The day after this photograph was taken the young ventured away from the nest. It is a move that cannot be delayed because barn swallows migrate and when the time comes to move out, the young of the year must be ready to keep up with the southward flight, and to feed themselves along the way.

Life Styles

The word *family* is usually defined to include parents, siblings (whether littermates or not), grandparents and other blood kin. In the animal kingdom that is too limiting a concept. A baby animal's family includes all the animals of its own kind with which it interacts.

Interaction, like *family*, is a word that must be accorded a broad definition if it is to be applied to the incredibly diverse spectrum of the animal kingdom. A single wildebeest calf may be one of tens of thousands born on a single day in a single herd in Tanzania. That newborn's family also encompasses the calf that is born at the same moment on the far side of the vast migrating herd. The two young animals make the same demands on their mutual habitat, are subject to the same predators, and may one day compete for the same mate. Herd, to herding animals, is family. In ways we do not understand, schools are families to fish, and pods to whales.

Some family units, of course, are more easily defined. Monkeys and apes interact endlessly in well-defined ways. Even very young primates interact with individuals other than their mothers. They play with siblings as well as with other young with whom their biological relationship is extremely slight. There is social grooming, there is the unquestionable authority of the mature male (including the young primate's true biological father), and the warning sounds of a distantly related cousin in time of danger.

Wolves care for each other's cubs, elephants establish protective shields for all calves in the herd, penguins and flamingos have nurseries with baby-sitters. All the things that human families do for their own young are done in the wild by units of animals, sometimes related by blood and sometimes not. To understand the concept of *family* we must think in much broader terms than we are accustomed to doing.

A young Japanese macaque, most likely this female's infant of the previous year, rests its head in obvious pleasure. The current infant cuddles and nurses on the right. This is family in the truest sense, animals exchanging security, animals needing and being needed and responding to the needs of others. This, surely, is where human society began.

A puma mother and her cubs. She will feed them, guide them, train them, and die protecting them if necessary. From the time her babies are born until they are ready to make it on their own, the female mountain lion is as tied to her young as the primate mother is to hers. Young, the only purpose of this family unit, are her full-time occupation.

These young opossum, members of one litter, need constant physical contact, not just for transportation but for reassurance and the building of confidence all mammals need. An animal without confidence may fail to survive, unable to make a decisive move that would favor survival.

Adélie penguin chicks are preyed upon by large gull-like birds called skuas. For safety the young penguins huddle together in their Antarctic nursery while their parents go to sea after weeks of fasting. The one white-breasted adult in the foreground and another in the background are in charge of the crèche. This is extended family with communal responsibility. It is an advanced social mechanism.

In Samburu National Park in central Kenya elephants crowd together shading the babies from the harsh sun. Providing shade is an essential activity in elephant society, where the family encompasses "aunties" and older siblings as well as mothers with their young. Small elephants in a forest of adult elephant legs are as safe as any young animal can be.

The raising of a litter of young lion ruffians is a tiresome affair and it is common for two or more females with young in tow to be found together. Feeding a cluster of young lions is more than any single cat could manage. There is safety in their numbers and vastly improved hunting efficiency.

Mountain goats (overleaf) are social animals and, except for solitary males, small herds are the rule. This social grouping, in a sense an extended family, seems essential to the animals' well-being.

The formation of herds, or pods, among whales remains something of a mystery. Belugas are sometimes found in both all-male and all-female herds, yet they have been sighted in mixed herds of up to ten thousand animals. Whether that many animals really form a herd or whether such huge congregations represent a temporary coming-together of many herds is not known. Nor is it known whether there are herd leaders or a hierarchy. Our experience with terrestrial animals would incline us to think that some kind of leadership exists, but thus far we have been unable to find any evidence of it.

The wolf is one of the most maligned of all animals, almost the exact opposite of what its reputation describes it to be. Wolf parents are the very models of family life. They protect their young in deep and well-padded dens, hunt for them, discipline them, and teach them how to survive in a hostile world. When the parents go off on a hunt, they usually leave their cubs with an unmated pack member who serves as a baby-sitter. When we refer to sexually promiscuous human males as wolves, we are again off the mark, for all wolves — males and females — are family-oriented.

Some fish school, while others appear to live an almost solitary life. The mechanisms involved in schooling are anything but clear, and any attachment one fish may feel for another is not only unknown to us but problematical. We can postulate, however, that schooling fish instinctively are most comfortable in close proximity to others of their kind and that schooling enhances species survival.

Colonial nesting terns on their eggs (overleaf one). In colonies such as these, thousands of chicks all appear at very nearly the same time. Isolated birds hatching over a period of weeks would suffer much more from predators than a dense clump like the one seen here. This is an almost unlimited extension of family, providing maximum safety and near maximum utilization of habitat.

An incredible congregation of Pacific walrus (overleaf two). Young walrus are born in a rookery, on an extremely crowded beach, and never seem to lose the need to be in close physical contact with others of their own kind.

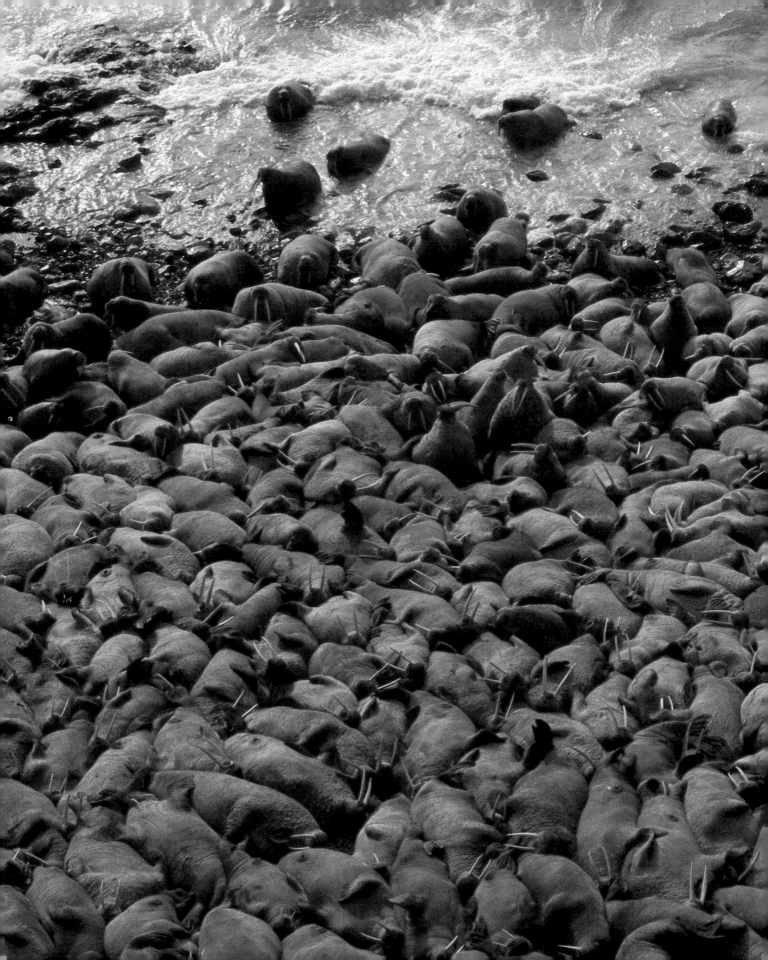

On Their Own

There comes a time when all young animals must go off on their own. Among some animals, bears for example, that means the young must literally move off and establish a new territory. Mountain lion young and jaguar young are expected to establish their own hunting areas away from their parents, for whom they are soon to be competition.

In herd animals *moving off* may mean little more than to stop nursing and start moving throughout the herd without the comfort of mother nearby.

It is difficult to say when family-forming species stop recognizing their young or the young their parents. At some point, though, that does happen. A young bison bull becomes a threat to the father as it reaches sexual maturity, and the father will drive it off, even kill it if necessary. A bear sow that is wildly protective of her young one month will drive them away with furious charges the next.

How much trauma an animal feels when it is finally rejected is hard to say. We know that in higher animals there is some. Young bears follow after their mother and her new lover bawling piteously until convinced in the harshest possible terms that the other side of the hill is safer. Many other young follow parents after there is no food or care left for them. It does come to all, however, all the finest and luckiest, a time of maturing when whole new facets of self must be discovered so that the process of procreation may begin again.

Immature roseate spoonbills and ibis fish in an alligator hole in the Everglades. The time has come for them. No parent will return to their nest with food offerings. Ready or not, the young are on their own. Only the very good have come this far. Only the best will go beyond.

The time when a young animal is on its own for good varies greatly. It may be years for an elephant, months for a large bird, but for the Ridley sea turtle it is moments after hatching.

Young polar bears such as these (overleaf one) are able to find their own food and will soon be off on their own into the vastness of their Arctic world.

For male elephants (overleaf two) to move off on their own is to link up with a grouping of their own sex. The females will stay with the herd, helping to rear the young and learning the rules of baby care until they, too, are ready to mate.

A puma (overleaf three), nearly full grown but still showing some of the spots of childhood, has become too large for its mother to feed and its presence in her hunting territory may be straining the food supply. It must head off and establish a territory of its own.

Snow geese (overleaf four) move south against the setting sun. Setting, too, is the vast complex of instincts that brought the young birds through their first summer and prepared them for their migration to winter feeding grounds. Although they will fly in formation and move as a group, each is very much on its own. No other bird will, or indeed can, stop to help in any way.

Text

New Life

Life begins with a minute egg, fertilized either parthenogenetically—that is, without the aid of a male—or by the male's sperm locating and invading the female's egg. When the nuclei of the sperm and the egg fuse, a zygote is formed and cell division begins; a potential new life is on the way.

For that egg, there is safety during its explosive development. The temperature is always right, nutrients are always at hand, all demands other than the single command to grow and develop are in the future. But this period of development is only a brief stage. Soon there will be birth—or hatching.

The birth or hatching of an animal is a traumatic event, certainly for the young and frequently for at least one parent. It is also a time of great peril accompanied by the challenge of setting the new, young lives into high gear as quickly as possible. The urgency is a survival mechanism and those that do not respond perish—not unfortunately, because their genes would be inferior stuff to pass along beyond their own brief time.

Ostriches illustrate the story of typical, if very much outsized, precocial birds. The parents become restless as they hear faint, chipping sounds within their eggs. The imprisoned young have responded to a set schedule of development and must hatch at approximately the same time, for within twenty-four hours after the first chick appears the parents will lead their family away and the nest site will be deserted. The dimly perceived noises of their clutchmates create a frenzy of activity among the still-unhatched ostrich chicks. For each there is now an imperative—escape! The chirpings of its siblings must be responded to or the chick will be left behind.

The ostrich chick's beak, temporarily equipped with a single, strong egg tooth, starts chipping through the shell. Breaking the inner membrane is easy, and although the shell of a freshly laid ostrich egg can withstand the blows of a hammer, it is dry and brittle as hatching time approaches because moisture has been robbed from it by the growing embryo. A chick that warrants

Ring-necked pheasant eggs hatching

survival will have the strength to break free; it takes only a tiny hole to breech the security of the egg.

The chick takes its first breath of hot, stale air from under the protective feathers of a parent. The shock of this initial contact with the outside world acts as a spur in the effort to be free. The beak has done what it can; now the powerful neck and legs must take over. There is a monumental thrusting effort and the shell is broken. The warm, dark, moist and, above all, protected world of the embryo is destroyed in the act of hatching. Life outside the egg has begun. Being hatched or born may be the toughest task any animal ever has to face. At no other time will it be so helpless and so open to disaster as it is during those first few moments.

The ostrich adults will lead their young and show them where and how to find food and water. They will shelter them from the sun and will brood them at night to keep the chill air from robbing them of body heat. But, of twenty or more clutchmates, only a handful will survive to maturity. External and internal parasites will rob them of strength. Jackals, eagles, and hyena as well as snakes and other predators will devour some; disease and accidents will take their toll of others. No animal parent can protect its off-spring from parasites and disease organisms. Mammals nursing their young may provide some early protection in their first milk but this means of immunity doesn't last long.

This, then, is the reality of entry into the world that overrides all others: there are no guarantees, no certainties, and only a chance of survival.

On the other side of the planet, in a North American forest, a white-tailed deer is about to give birth. It has been seven months since she mated and now, gestation over, she has chosen an abandoned fence row as a site for the birth of her firstborn. As the time approaches she paces nervously. Then the tips of two small fore-hooves appear from the birth canal and slowly extend. The doe lies down gingerly, her discomfort obvious, licking the place that is causing her distress.

Ring-necked pheasant hatchlings

Shortly after his feet appear, the little buck enters the world with a thud, expelled from his mother's body by her powerful, violent muscle contractions. With his first breath he inhales her scent, a recognition of which is the essential first bit of knowledge he acquires. The mother then consumes the placenta and licks her newborn dry of amniotic fluid; the nutrients of the ingested placenta will nourish both her and her infant during the thirty days, at least, that she must nurse it. If predators, parasites, disease, natural disasters, and *man* do not prevail, the young deer can be expected to live ten to fifteen years and establish a territory of its own not too far from where he was born.

Within hours of cleaning her new fawn, the doe will nudge it to its feet and move it to a more secure area, away from the strong odor of birth that would be certain to attract predators.

Newborn Jackson's hartebeest

A white-tailed fawn is programmed by instinct to remain hidden while its mother is away browsing. The doe's food demands are heavy, for now she must provide milk for the fawn and seek nourishment for herself. She returns several times during each twenty-four hour period to suckle her fawn, and as the youngster nurses, the doe licks its genitals and anus to stimulate urination and defecation. Once again she controls scents that could give her baby's hiding place away. She cleans the waste material from the fawn and from surrounding ground litter.

The fawn has no scent of its own at this early age and thus is at least partially protected from predators. When its mother is gone, it will instinctively "freeze" if its senses an alien presence. A would-be predator, unable to distinguish color, will usually walk by the inert and odorless form, mistaking its spotted baby coat for part of the natural environment.

Not all baby animals receive parental care, however. Many species come into the world with a fixed pattern of behavior that guides them through life from its beginning. Almost always, these young are produced in large numbers, so it does not affect their species if only a few survive. Attrition is allowed for in advance.

Milk snake with eggs

For example, it would be impossible for a female sockeye salmon to protect the ten-to-thirty thousand eggs she will lay in a single day. The swift current of the gravel-bottomed salmon stream will claim some; others will be taken by predators, while still others will fall victim to fungus invasions. A single brown bear, placing a large paw into the water while drinking, can crush or send into oblivion thousands of potential salmon young. This possible decimation has also been allowed for in advance. It is unavoidable and anticipated.

For many creatures, such as salmon, the innate response patterns that protect them make them akin to automatons; their teachers are sunlight, temperature, and the chemical composition of their natal waters. If all goes well, a salmon will pass through the stages of embryo and babyhood, with instinct and form perfectly matched to environment and demand, until as an adult it heads to sea and a whole new set of demands, chemical as well as physical.

The salmon may spend as many as seven years in the sea before body signals order it home to its ancestral river to produce young. It is a foolproof method unless a sudden change such as pollution, occurs. Then there can be no redress. The cycle simply ends. Interestingly enough, the Atlantic salmon repeat the cycle year after year, making many trips into and home from the sea. The Pacific salmon, although descended from the Atlantic species, have made a strange adaptation. They return only once to breed and then die within hours.

Except in obvious details, the life of the eastern box turtle is similar to that of the salmon. The female, some time after reaching her fifth year, encounters a male who advertises his availability through scent signals. After a brief courtship the male mounts the female, his bottom shell being adapted to permit the unlikely performance. The sperm received from the male will remain alive within the female for years, allowing her to produce fertile eggs for several seasons without coming in contact again with a male. This system is, in part, a response to low-population density.

In the early summer the female will search for a nesting site. She will automatically seek a well-drained area, not susceptible to flooding, which will receive the sun's warm rays. Only a certain amount of sunlight is required, however; too much would bake the earth into a brick-hard prison for her unhatched young. She has had no guidance; she was taught nothing. Yet instincts guide her to exactly the right place. The very fact that she has survived is proof that she was born with her instinct package intact.

When the box turtle has found a likely spot to lay her eggs, she urinates to dampen the soil and facilitate digging. Then, in ponderous turtle fashion, she excavates a flask-shaped hole. It takes her several hours, as she must do the digging with her back legs. Whe she is done she deposits up to eight elliptical eggs in the cavity and covers them with the loose soil she has dug out of the hole. Before leaving, she hides the site of her nest by disturbing its surface to make it indistinguishable from the surrounding area. That is her final act as a mother. The eggs are no longer her concern. She has fulfilled her maternal purpose, at least for the year.

Should the young box turtles not hatch until cool temperatures have arrived, they will automatically know enough to hibernate rather than leave their protected underground home. They will spend the winter in their fractured shells even though they have actually broken free and could reach the surface.

On the average, of the eight eggs produced by the female, only six will hatch. Perhaps three of the six hatchlings will be taken by fox, raccoon, crows, feral dogs, children, or automobiles. If three survive to reach maturity and themselves become progenitors of another generation, it will have been a successful brood.

The aphid's life cycle starts with eggs laid in the fall. The eggs overwinter and hatch in the spring. The new young will all be female and they will give birth to successive generations of females parthenogenetically, without being fertilized by a male. Over a dozen such generations will be born alive in one summer. Then, as fall approaches, males will suddenly be produced within the colony.

Newborn white-footed mice

135

These males mate with females produced at the same time; they, in turn, lay one generation of fertilized eggs that will survive the coming winter to hatch the following spring as females, when the cycle will repeat itself.

Because there is no parental guidance, these insects must rely on internal signals so complex they may never be fully understood by man. How, for example, does the aphid know that it is time to produce males, and how is the signal communicated to the reproductive system? The aphid, of course, never "knows." The reproductive system receives its own signals, probably from hours of sunlight and other seasonal readings.

No catalog can list the different styles and variables of birth. In our discussion we can only seek the denominators common to the most species, those that best illustrate the incredible range of devices, systems, mechanisms, and subtle alterations that have evolved in order to allow life to take full advantage of every scrap of inhabitable environment there is. Some life is very adaptable and can handle a wide range of conditions—many species of rodents are good examples of this adaptability—while other kinds of animals are severely restricted. The Everglades kite is a bird dependent upon a food source so limited that even minor environmental changes can spell species disaster.

A young animal emerging into the world may be able to handle a small range of conditions, but for most species the ability of the newborn to adapt is extremely limited. Most animals are born or hatched to tolerate conditions similar to those their parents found, as did their parents before them. If things are not too different, the young animals, newly emerged from womb or egg, are ready to embark on the next stages of their species plan. Style and techniques may vary, but each will emerge with all systems tuned and ready to begin a new generation of life, able to do what inanimate substances cannot do—draw on the environment, use it, and then reproduce itself in kind. The cycle of birth, reproduction, maturity, and death is the story of all living matter.

Thirteen-minute-old white-tailed fawn

First Days

The most dangerous period for any animal is the time between birth and early maturity. Genetic vigor and natural selection have been preparing and equipping this new life for millennia, but the forces that would destroy it have also been long in the making. Even its genetic and selective protection may work against a young animal if it does not conform to the norm. (Childhood is a time when improvisations within a species may appear, but few such individuals will be adaptive enough to survive and pass on their new tools for survival.) Natural selection even pits sibling against sibling. In times of short food supply, only the strongest can survive and this is achieved at the cost of the lives of their littermates.

What of the young songbird whose pre-cast lot places it in a nest which will be raided by the parasitic cuckoo? The female cuckoo does not build her own nest, but lays her eggs in the nest of other species. She may even throw the legitimate eggs out of the nest before laying her own. If the unwitting songbird mother should lay additional eggs, the cuckoo hatchling will force the other eggs and chicks out of the nest so that it will get all the food.

For those species which have highly developed senses, birth or hatching must be accompanied by a feeling of fear. Wrenched from the security of the egg or womb, all that it sees or hears must seem life-threatening. In those species where instinct rules, terror is most likely unknown, replaced by a single-minded purpose to seek shelter from predators and food in order to continue the growth process.

Some species have developed a method called "imprinting" to calm these early fears and create a role-model which will guide and nurture the youngster to adulthood. The eminent ethologist, Konrad Lorenz, is perhaps most closely identified with the process of imprinting. This, in simple terms, is an instinctive ability of an animal to identify itself with the first thing that it sees after birth or hatching.

A mammal or bird that is meant to have a family must be able to identify that family as its own as quickly as possible. Therefore,

Great horned owls in nest tree

Nursing opossums

Grizzly cubs

the young animal is programmed to relate to the first mature animal it encounters. This mechanism is particularly important in waterfowl and enables large broods of youngsters to remain in contact with adults as well as siblings.

There is a seemingly endless list of diseases, accidents, and natural disasters which may befall the newborn of any species. Many parasites and diseases tend to be more stressful to the very young and very old.

Natural conditions must be close to optimum for the survival of young. Untimely storms or periods of intense heat or drought may severely limit the number of survivors in a given season. Populations, in fact, may differ dramatically from year to year because of weather variables.

The inexperience of adults may play a negative role in the survival of infants. The first young of cats and primates have a much lower survival rate than successive litters. It appears that this may be a learning process for the female to better enable her to cope with future young. Studies have shown that some species of doves are more adept at nest-building with each succeeding attempt.

Man (and his technologies) makes an impact on each new generation of nearly every species. He may even inhibit adults from breeding by destroying critical habitats. His poisons may lead to sterility in either predator or prey species which make up an ecosystem.

The severely endangered peregrine falcon and the recovering brown pelican have fallen prey to DDT, one of man's more diabolical poisons. While these birds may live and breed successfully, the DDT in their systems may inhibit proper calcification of their shells. This means that the embryo growing inside may die when the soft-shelled egg breaks prior to hatching.

Although the use of DDT is strictly monitored by the United States government, it persists for a long period in soil and water. Its effects are not yet fully known, but they have similarly affected the bird that is our national emblem, the bald eagle. The southern bald eagle's rapidly decreasing habitat is only a partial reason for its decline; the effects of DDT are constantly weakening its delicate hold on life.

DDT is absorbed by the bodies of microscopic plant and animal life. These tiny forms are then eaten by larger species and so on up the food chain. Fish are surrounded by food sources containing DDT and feed on them constantly. This causes the concentrations of DDT in some fish to be higher than that of the lower organisms. By eating fish, the eagle is at the top of its particular food chain and is therefore even more affected by the DDT. Man is also at the top of his food chain.

In the remote Greenland Sea sits Jan Mayen Island where the "Greenland herd" of harp seals has held its annual "birthing" for centuries. The migratory force within the female harp seals steer them, along with others of their kind, to this normally deserted area. The females are the first to arrive and from mid-February to early March give birth to their young. Newborn harp seals will stay on land only from two to three weeks before heading out to sea as young adults to seek their own fate, and even with no interference this is their time of extreme vulnerability. But their downfall is the long, white, silky natal coat which they wear until their more permanent pelage appears. Man, in his quest for the expensive, the exotic, and some believe, the decadent, covets this fur.

Infant red squirrels

Few baby animals are brought forth into a more hostile environment than the harp seal. The steaming baby emerges helpless onto the glaring ice, and for the next two weeks is totally dependent on the protein and fat-rich milk of its mother.

It is during this period that scores of intinerant workers and people simply out for a spring lark, armed with clubs, move through the colony creating havoc. Innocent, sloe-eyed harp seal pups look up at the invaders, wholly unaware of what is coming. With each crash of the stunning club, another life is sacrificed in the name of fashion. Some are killed instantly; others are not so lucky. Some feign death instinctively as a survival tool.

Five-week-old bobcat kittens

Raccoons in tree den

Harp seal colonies now number less than one-third of their original strength and the annual harvest rate is so high that it is apparent that even natural selection cannot reverse the inevitable. Environmentalists believe that too many of the young seals are being killed to allow for regeneration of numbers in the species.

The animal family of lagomorphs contains hares, rabbits, and pika. Although for many years, these animals were classified as rodents, this had more to do with their similar feeding habits and their continually growing incisors than with anatomy or line of descent. In this family, the hare and the rabbit have developed very different methods of protecting their vulnerable young.

The true hares are more widely distributed than rabbits and differ from them in many important ways. Hares, as a rule, are much faster than rabbits. The hares choose their nest sites aboveground, and the young, being precocial (able to follow the parents soon after birth), flee the nest when frightened.

The hares' dens are built in dense cover and are scraped-out depressions in the soil. A female may have more than one den site and move from one to the other to confuse predators.

Hares have a "flight distance" of about three meters and will "sit tight" until that space is breeched. Should that territory be invaded, they literally erupt from the den and take rapid flight.

After a gestation period of about forty-two days, one to six young are born, depending on the age and condition of the female. The young nurse for two to three weeks before being independent of the adults. Young hares are born with teeth, are sighted, and covered with fur, appearing much like miniature adults with the exception of somewhat foreshortened snouts.

These fully developed youngsters are in direct contrast to the young of the closely related rabbits. Rabbits are born blind and without hearing or fur. They are altricial—that is, they are helpless and unable to leave the nest for the first few weeks of life. By four weeks of age young rabbits are completely independent, and their mother is preparing for her next litter.

Young rabbits, because of their vulnerability, are born in a very different setting than the fairly open nest sites of hares. Rabbits prefer to dig elaborate underground burrows. For this reason, rabbits are most often found in areas of sandy or loose soil. These underground warrens may contain several entrances and blind passages to thwart predators.

So, although closely related, rabbits and hares deal with the problems of raising young by two quite distinctly different methods.

Rabbits and hares do, however, share one curious and beneficial trait which is peculiar to them. Rabbit feces are generally hard, dark pellets. Supplemental fecal material is also produced, and these fecal pellets are generally softer and covered with a mucous layer. These are re-ingested by the animal and are stored in a separate section of the stomach for further digestion. These special pellets enable the rabbit to extract nutrients which were not broken down by initial digestion. This is very similar to the practice of cud-chewing in ruminants. The pellets are rich in vitamin B1, and allow the animal to survive during times of reduced food supplies. It has even been proven that rabbits kept in captivity on wire, where the fecal pellets drop through for ease of cleaning, do not thrive. This because they are unable to re-ingest their vitamin-rich pellets.

Red-shouldered hawk chicks

Birds also must deal with a variety of hazards while raising their fragile young. Passerine or songbird young face a very different set of circumstances at hatching than, for instance, the young of waterfowl such as the Canada goose.

Like rabbits and hares, the young of these two families have evolved along different behavorial lines to enhance the survival of their species. There are two basic types of young birds: nidifugous and nidicolous. The nidifugous young are able to follow the adult bird, leaving the nest and searching for food soon after hatching. The nidicolous bird is born naked and remains in the nest for an extended period of time while being cared for by the adults.

The American robin is a good example of nidicolous young. Young robins are born in a cup-shaped mud-and-straw nest which is familiar not only to the amateur naturalist, but to most children. When the sky-blue eggs hatch, it is a frantic time for the harried parents, trying to keep their young fed.

The first task of the parents is to remove the bits of eggshell and either ingest them or drop them away from the nest so as not to attract predators. Then comes the wearying task of searching for animal food to feed their eternally hungry youngsters. Although the young are born blind, the movement of the nest when the adult alights will waken them and signal them to open their large mouths to beg for food. This process goes on from dawn to dusk. After eating, the young will defecate almost immediately, and the adults will either ingest the fecal sac or throw it outside the nest.

This laborious process will go on for a short time after the young are fully fledged, because even after they have left the nest, they are not yet fully able to procure their own food.

The Canada goose has evolved a different mechanism for raising her young. The young of the Canada goose are nidifugous.

The Canada goose nesting process begins in early spring, following a long migration to the muskeg and Arctic tundra. Soon after arrival, the mated pairs stake out nesting sites and construct their nests. Because of the short Arctic summer, everything must be on schedule or the young of the year will not be ready for the rigorous return flight to the winter feeding areas farther south.

The clutch of eggs generally numbers about five, and hatches in an average of twenty-eight days. Depending on latitude, the young must be fully fledged and ready for flight somewhere between forty to eighty days after hatching. Those who cannot meet this rigid schedule are destined to die before their first migration.

The average clutch requires about a day to hatch, and before they can follow their mother, the young must be sufficiently dry to be waterproof. The first night, then, is spent in the nest. The female may brood youngsters for the next few nights as they strengthen.

Three-day-old white rhinoceros

During those early critical days, both adults lead the youngsters to feeding sites and valiantly protect them from predators. Each succeeding day sees the young become more independent as they prepare for the fall migration.

This stressful time is extremely important for the continuation of the species. Obviously the young must survive, but many other processes are taking place during this period. These begin before mating in many species in which an extended pair bonding is essential to insure that specimens of the same species will mate with each other and that they will stay together during the rearing process to foster the growth of young.

In species such as those of the lion, the chimpanzee, and gorilla, a bond must also form between the mated pair and the other members of the group of which their offspring will someday be an integral part.

Fur seal pup

Many animal societies have rigid codes of behavior, and failure to adapt to them can end in death. This ultimate punishment may come from the group leader or, more often, from a predator. Primates, raised tragically as pets or hand-reared in zoos, can rarely be placed back into a natural breeding group. They have not been given the important training by parents, siblings, and other group members. They may be killed or spend the rest of their days as social outcasts cowering in a corner, afraid to move, because they don't understand the ramifications of their actions.

Childhood is a harsh time for any wild animal and a constant drain on adults. But the natural world is not cold and uncaring. Anyone who has witnessed the birth of a wild gorilla has been almost certainly amazed by the reaction of the group. When the youngster is first shown to the troop, there is great excitement, with crowding and jostling to see the baby. The troop members babble and stroke each other for reassurance; the mother immediately rises to the top of the troop's social rank. It is a time of joy in the group—rebirth, succession of the species, new life!

Parental Care

Parental care is amazingly variable in the animal kingdom. Compare the attention required by a human child with that needed by the young of the many species who survive perfectly well without parental guidance.

Those species that receive no parental care are programmed to survive by innate behavior. Their instincts protect and drive them on to find food, shelter, and eventually mates. They are often protected by massive numbers that insure that some will survive to procreate. Some species—certain spiders, for example—are cannibalistic, and fewer young spiders emerge from the egg cases than eggs were laid.

We have already seen that the salmon would be powerless to provide guidance or food to her thousands of offspring. The voracious young of the snapping turtle leave their underground nest and head to the nearest water. No parent is needed to tell those powerful jaws what is food and what isn't. Almost everything is!

Although there are some variations, most snakes, whether born alive or hatched from eggs, head off into their world with only inborn skills. The myriad insects produced in the world are left mainly to their own devices, as are many spiders. (Exceptions, of course, are the highly social insects: termites, ants, and many wasps and bees.)

Limited parental care seems to work well for a number of species. The American alligator amasses a huge pile of vegetation in which to lay anywhere from twenty-five to sixty eggs. After the nest is completed, she hollows out the center, lays her eggs, and waits for the combined action of rotting vegetation and sun to hatch them. When the young make contact with the female by chirping, she uncovers them and leads them to water.

The female wolf spider carries her youngsters with her as she seeks prey in much the same way our own North American wolf hunts; hence her name. Her eggs are contained in a silken sac attached to the spinnerets which spin the protective casing. The sac is dragged along behind her until the eggs hatch. The young wolf

Mute swan cygnet on mother's back

spiders then crawl up onto the female's back and are carried around for about a week until they are allowed to fend for themselves.

The king cobra is one of the few snakes which does provide some parental care for her young. She constructs a nest by gathering material with her coils. The nests have a two-story construction, and the lower area contains up to forty hard-shelled white eggs. She may wrap herself around the nest, but does not truly incubate the eggs. It seems all her energy is saved for protecting the nest, and she will take on any intruder who dares threaten the security of her developing eggs. An incubating king cobra is greatly feared in its natural environment. The species is the largest poisonous snake in the world, reaching a length of almost six meters.

Some species of python, although not as fiercely protective as the king cobra, are actually able to incubate their eggs. The eggs, which are often clumped together in masses, are surrounded by the female. She is able to raise her body temperature by making convulsive moments resembling shivering. The pythons' parental responsibilities are completed upon the hatching of the eggs.

Nursing lion cubs

Obviously, many species provide highly specialized and time-consuming care to their young. In such species, the responsibility begins at the moment of birth. The helpless young of these animals could not survive the first day without the help of the adults.

The gestation period of the African elephant is nearly two years. The development of the young within the group roughly parallels that of human growth in society, including those difficult teenage years. The young of the chimpanzee, even if they were able to become self-sufficient, could never be a part of a social group without the elaborate training provided by parents and other group members.

Few animals are more vulnerable than a newborn African lion. Without parental care and feeding, as well as behavioral learning, it could never become a part of the pride that is so essential to its survival and the survival of its species.

Even the zebra, which, like most of the horse-related species, is not thought to be overly intelligent, is a part of a highly structured family group which protects it into the next generation.

The parental bond or, more importantly, the mother-infant bond is the first step in socialization for those species which require group participation. This bond introduces the youngster to the group and makes it more adaptable to the other important bonds yet to come. The second step is the bond which makes the youngster a part of the group to which it belongs. This is the bond that will allow the young animal a part of the group's bounty, at least in the case of predators. This same bond, in many species, will make it responsible to help feed and protect the other members of the group when it reaches adulthood. Finally, in some species, there is a pair bond. This may only last for the length of copulation or may be protracted to include the care, feeding, and raising of the next generation. For some few animals it lasts a lifetime.

The birth of a zebra on East Africa's plains generally coincides with the rainy season. This provides lush grazing for the lactating female and later for the growing youngsters.

Immediately prior to birth, the zebra mare moves away from the family group, which generally numbers seven to ten individuals. The group remains a short distance away, but the stallion stays near the pregnant mare. This is a time of extreme vulnerability, and he stands watch, ready to react to and possibly even challenge predators.

Birth must come quickly for obvious reasons of safety. Predators are always nearby. Miniature hooves appear first, then the nose, and later the head. Midway, the mare will lie down. With umbilicus still attached and while fighting to be free of the amniotic sac, the foal sees the first light of day.

With all the effort it can muster, the foal breaks free of the foetal sac and crawls toward the mare's head. Its mother will lick it and clean it of fluid, ingesting the material as she proceeds. This

Burchell's zebra nursing

Nursing sea lion

146

serves an important function: distinctive smells cement the mother/infant bond.

The mare then stands and waits for her foal to gain strength. Within an hour, the foal is able to follow its mother. Many plains ungulates do not have the luxury of so much time to get their young to their feet and moving. But the strong social grouping of zebra families allows the extra time.

The newborn foal's legs appear much too long, and the beginnings of the mane are only scattered patches. Even its color is wrong; but in the weeks following birth the brownish markings of the foal will change to the familiar black-and-white striped pattern. Most of us would describe the zebra as white with black stripes. The Africans, however, tend to describe them as black with white stripes. It does tell us something about perspective.

Nursing begins within the first hour, and the mare during this period will chase away any other female. The pair-bond has not yet been fully established, and the mare wants to be sure her foal follows only her.

Foals, like other young animals, learn many important lessons while playing within their group, with other foals at first, and later with the adult members of the group. This play teaches alertness and cements the bonding within the group.

The foals are weaned at about four months, but continue to follow the mare for up to two years. During this time, the mare may have produced two more foals. All her young will continue to follow her, the more mature bringing up the rear. And, of course, at the extreme rear of any zebra group, the stallion is jealously guarding his mares as well as keeping a wary eye out for predators.

The parental guidance during the two-year maturation process is essential to the survival of the foal. It must be accepted by the group and must understand all the signals which will protect it during its life within the herd.

The North American whooping crane is one of the most

Alaskan brown bear sow with cubs

Cow and calf moose

Rocky Mountain nanny goat and kid

seriously endangered birds in the world. Although there are now over one hundred birds in the wild and in captivity, in 1941 their numbers had dwindled to fourteen. This caused great alarm, and conservationists were able to convince the government to afford full protection to the species. Nearly twenty-five years later, the nesting area of the whooping crane was found in Wood Buffalo Park in Canada, and a massive campaign was begun to save the species.

This process has given scientists a rare opportunity to study parental care and its effect on offspring. When the nesting grounds were surveyed, it was found that, although most crane pairs laid two eggs, only one would reach maturity. So scientists made plans to remove one egg from each nest and bring them back to the United States for captive propagation. It was hoped that this would provide a breeding population for later transplants with the more populous sandhill crane.

In 1975, fourteen whooper eggs were collected in Canada and used to replace sandhill crane eggs in nests in the Grays Lake population. This was a gamble with very precious eggs to see if sandhill cranes could foster-rear the young of their more endangered relatives.

The transplant was an initial success, and six young whoopers survived the first season. It was strange to see adult sandhill cranes leading their much larger foster offspring, but it appeared the bold move might be successful.

Of course, the process was not without problems for the foster parents. When the foster parent sandhills would sound the alarm call, the whooper youngsters would head for deeper water, a trait common to their species. However, for safety, sandhill chicks would have headed with their parents to tall grass. This caused more than a little frustration for the sandhill parents and, along with other innate whooper responses that the sandhills could not anticipate, might well have caused chick mortality.

A number did survive, however, and the program continues to

be successful. But the acid test looms nearer and nearer. Will the whooper foster young attempt to mate with whooping cranes at maturity or with the smaller sandhill cranes from whose ranks the foster parents came? The opportunity for scientists to monitor this program has given us a rare insight into the nuances of parental care. Will instinct win out, or will the tutelage of foster parents create dangerously mismatched pairs of cranes?

The African elephant may well have the most highly organized social unit of any mammal. The species depends on the support and guidance of all the members of the herd to see that the young elephants make it through their long maturation process of up to twenty years. The sub-group, into which a young elephant is born, is entirely matriarchal, and the males in the group are those who have not reached sexual maturity and who have not been relegated to the bull groupings.

Uganda kobs

The infant elephant is totally dependent on its mother. As a matter of fact, although it will play with other calves, it will be nearly two years old before other adults will take an active part in its socialization.

Directly after birth, the mother elephant kneels slightly to help her calf take milk from her amazingly humanlike breasts. The calf's trunk, at this point, is totally useless and will be used later only after a great deal of practice and strengthening of the muscles. The female constantly caresses her youngster and keeps it very close to reinforce the essential bond. At the slightest alarm, the infant is herded underneath its mother's massive body for protection. In extremely hot weather, her body will shade her youngster, and if a stream must be crossed, the baby will always be positioned upstream from her, secure from any currents or eddies.

At the water hole, the mother elephant sees that her youngster does not become mired, and she carefully washes it with water from her own trunk. Almost immediately, she covers it with dirt to protect it from insects and sun.

At six months, the baby elephant may play more in calf

groups, but it is never far from its mother. The activity within the play group provides valuable lessons for the future.

When her baby reaches two years, the cow tends to give her youngster more freedom. Older siblings and aunts within the herd continue the teaching process. But those first two most important years are the private domain of the mother.

We know that many primates cannot raise their own young unless they have seen their mother raise a sibling. It is difficult to imagine a totally isolated elephant cow trying to raise her calf without the lessons she learned from her mother, aunts, and siblings.

Parenting is symbolic of society. No society can exist without strong role models of what is right and what is wrong. Bad decisions and bad reactions in an animal society can lead to destruction, not only of the offender, but of the group. Group membership brings responsibility, and responsibility cannot be innate. It must be learned; different lessons must be repeated, and examples must be given.

The parent is the transition from self to the social unit. In this way, animal parents are no less important than human parents. They are teachers, and the lessons are difficult; but a good grade equals survival. The social grouping is obviously evolving as a dominant force in nature, and the role of the traditional parent can only grow in those higher species allowed to survive the terrible onslaught of man.

European lynxes

Growing Up

After the shock of birth or hatching, the learning of new behavioral patterns marks the next stage of development. The process of growing up is a difficult one.

For some animals, growing up is not as difficult as it is for others. The salmon moves through its many stages of development simply by eating and following inborn signals that lead it through its life of reflex, response, and minimum learning. The cockroach is spared the sometimes difficult lessons that must be learned by, say, African lion cubs. The roach merely seeks warm, moist darkness and sufficient food. The growth process for the roach is a series of molts or sheddings until adulthood has been reached. But nothing is really that simple in the animal kingdom. The price that many of these less socially oriented animals pay is individual survival. The success of their species is dependent on multitudes of offspring, and this method ordains that only a few individuals may survive of the many produced.

Mammals and birds generally have structured juvenile periods. There are many lessons to be learned during this time. Baboons must teach their youngsters to deal with the potentially deadly scorpion that is also an important source of protein. But an incautious individual might succumb to its intended meal without the guidance of elders. The young prairie dog must master the calls and body postures which are its ancestral language. Mistaking a greeting call for a predator alarm may miff a fellow prairie dog, and could inhibit breeding. A mistake in the other direction may only have survival value for the alert red-tailed hawk.

We will see later that the growing-up process for chimpanzees is intense and difficult and takes place over a long period of time. The extent of the role the social grouping has in species life dictates the length and intensity of the juvenile period.

Young birds generally have two periods of development. The first is the period of total dependence on the parent birds for warmth and direct feeding. Even birds such as pheasants, wild turkeys, and waterfowl, which lead their young to food sources, must

Lion cub

Raccoon mother with two young

Hippopotamus and youngster

brood them until they are able to maintain their own body temperature. This period is more easily illustrated with members of the pigeon family, which feed their young a concentrated "pigeon milk," a product of their own crop. There is one dramatic exception in the bird family to this process and the one to follow. The mound birds or bush turkeys (they are *not* true turkeys) of New Guinea and Australia have evolved an interesting process of hatching young. Their eggs are laid in rotting masses of vegetation sometimes as large as twelve meters in diameter and five meters high. Some species mix lava or hot-spring-warmed soil in the mound. Others use beach sand, and the sun attends to the task of incubation. When the young of these species break out of their underground nests, they head for cover, and at this stage shun all living things, even their own parents. They are down-covered when they emerge but have fully developed flight feathers and are able to flutter to a branch within twenty-four hours. It is a strange, but successful, adaptation.

The second level of bird maturation is the stage when the young are independent of their parents as food sources but still need the family unit for protection and instruction in social group behavior. Waterfowl are good examples of this level of development.

Mammalian growth can be divided into three periods. The first is that of extreme dependence on the female for milk; this period is even more important in the case of carnivores, whose young are altricial and require relatively more care than ungulates.

The next phase is one of mixed dependence. The young animal is still receiving milk, but is beginning to take solid food. Its total dependence on the female is still apparent, but it is gradually being introduced to its coming independence.

The final stage finds the juvenile totally independent in the search for and selection of food, except in those social species where hunting is a group endeavor. This is a period when the youngster is sharpening its survival tools with the aid of parents. It is on its way to full independence.

Play is of vital importance at this third level. The young lion chasing its mother's tail is sharpening its senses for the time when success at hitting a moving target may spell the difference between an empty belly and a full one. The four-month-old bull elephant who mock-charges quaking leaves and dust clouds is learning valuable behavior for the day when he will assert himself.

The chimpanzee is one of the best loved of all wild animals. But what do we really know about chimpanzees? It's not generally known that an adult male chimpanzee may weigh nearly two hundred pounds, reach fifty years of age, and be three times as strong as a man. That's because the chimps we see at circuses and on TV are generally under five years of age. Once a chimp reaches sexual maturity, it can no longer be freely trusted in human society.

Behaviorist Jane Goodall has proved that chimps are capable of using tools. They poke sticks into termite mounds and pull them out covered with insects; these are licked off and the stick returned for more. In times of drought, chimpanzees will chew leaves until they form a makeshift sponge and insert the tool they have created into the hollows of trees where water has collected. They then wring the liquid into their mouths.

Great horned owlets

They may be ingenious for an animal by any standards, but chimps tested and studied in human environments have given even more spectacular insights. Thirty years ago, a pair of scientists raised a chimpanzee baby as a child in their household along with their own child. The chimp was taught to do many things which she greatly enjoyed. One involved presenting her with piles of photos, clippings, and snapshots; her task was to place the photos of chimpanzees in one pile and humans in the other. She could accomplish this with amazing accuracy. But she consistently made one error: she placed her own picture in the human pile.

The young chimp also went through a period of dragging her hand behind her with knuckles nearly touching the floor. She seemed self-conscious of this action and would do it only when she thought she wasn't being observed. Her surrogate mother equated this behavior with the human child's developmental period during which pull-toys play an important part.

One day the ape stopped abruptly and made tugging motions while running around the toilet in the family bathroom. She looked up at her "mother" imploringly. There was unspoken communication, and when the scientist made motions of untangling the imaginary toy, the chimp jumped to her arms with glee. So much for the theory that primates are not capable of abstract thought. We have now trained chimps to communicate by computer and with grammatical accuracy.

These examples obviously denote a very advanced form of animal life. Growing up in a chimpanzee troop in the wild is a life full of experiences which allow the young ape to enter one of the most complex animal societies on record.

The chimpanzee is born after an eight-month gestation period. If born to an experienced mother, it will probably have been three to four years since she last gave birth. If the young chimp has siblings, its life will be greatly enriched, for the siblings stay with their mother, sometimes into their teens, and act as playmates for younger brothers and sisters.

The baby chimp stays close to its mother for the first six months. She is extremely solicitous of her newborn, which clings to her belly fur as she travels. Between six and eighteen months it rides jockey-style on its mother's back, but will be nudged to her belly in times of danger.

At the point when it begins to ride like a jockey, the baby will begin to beg for solid food, which it receives directly from its mother's mouth. Taking food directly from its mother is a way of learning what is, and what is not, eatable. The youngster is now allowed more latitude and plays with siblings and other young chimps within the nursery group. A great deal of play takes place which will have significance later. Older chimp youngsters practice at "mothering" and carry and play-protect the younger members

Olive baboon

of the group. Rolling, biting play will help the young males in later dominance struggles. They practice loud branch-slapping charges and mock-threat displays which are vital in identifying dominant members of a group.

Chimpanzees have been found to have twenty-three distinct calls or grunts. The young learn the significance of each call through interplay with both siblings and adults. Facial expressions are also very important in chimpanzee society and must be learned, as a great deal of their communication is non-vocal. What would appear to be a grin to humans is actually a signal of abject fear in chimpanzees.

Grooming is an important part of many primate societies, and the chimpanzee is no exception. Hours may be spent in grooming sessions, which serve many purposes. The most important is to cement the bonds which hold the group together. More practically, precious salt is gleaned from dried skin.

The long period before adulthood in chimpanzees is indicative of a high social order. Lessons must be taught, feats attempted, and the nurture of the female solicited when bravado fails. These lessons take time, and their value to a youngster cannot be overestimated. There have been documented cases where females died and left seemingly independent young who thereafter failed to thrive. They eventually died without the support of the strong maternal/young bond.

The Couch's spadefoot toads are found in the Midwest, often in arid regions. These amphibians are at the mercy of sustained moisture and rainfall that create their breeding habitat.

They get their name, "spadefoot," from the shovel-like plate on their hind legs. With these built-in tools, the toads bury themselves in soft sand in order to keep their skins moist; they can bury themselves in loose soil in a few moments. Spadefoots are rarely seen in daylight, and their range is restricted by the consistency of the soil and its ability to retain moisture.

During periods of infrequent rainfall, multitudes of spadefoot

Two-month-old grizzly

Raccoon climbing

Lynx kitten

toads emerge and frantically begin their procreation process. All available adults head for temporary pools and run-off areas to mate and lay their eggs. The female can lay up to a thousand eggs in one long string.

Breeding takes place in one night: then the adults return to their underground sanctuaries and the young's race to survive, before the transient pool dries up, begins. The young will hatch within two days, live on their yolk supply for an additional day, and then begin to hunt on their own. The tadpole stage is rapidly accelerated due to the limited availability of water. The pools will only last for a few weeks at most, and the miniature toadlets must metamorphosize as quickly as possible.

The flea-sized toadlets have been known to reach adulthood in only twelve days and head for life on—or, more properly, under—land. This stage of growing up for the Couch's spadefoot toad is frenetic but essential for survival.

Growing up places tremendous pressures on both adults and their young. There is extreme hazard from predators. The period of vulnerability begins with mating and heightens during the pre-occupation of the female during birth.

Young animals are racing against time to reach the relative stability of adulthood. Young mammals may be hampered by "long bone disease" which simply means that the body is developing out of phase and some parts are growing more rapidly than others. This usually causes only minor discomfort and accentuates the youngsters' ungainliness.

The lessons young animals learn are vitally important, as studies in captivity have shown. Dr. Harry Harlow was a pioneer in infant-deprivation studies with primates, and proved that young primates deprived of adult females at an early age were barely able to function as adults and were generally incapable of breeding and rearing their own young.

Zoos have made efforts to reduce the numbers of young who must be hand-raised by human surrogate mothers. This is more essential with primates than with other animals. Some zoological institutions are experimenting with primate surrogate mothers. The primates chosen are females of closely related species. It has been proven that a major need of the infant, primate or human, is movement or "rocking." Deprived primate infants clasp themselves and "rock" in a pitiful attempt to provide for themselves the developmental stimulus they require.

Live surrogate mothers may not be able to lactate and feed young, but feeding can be done by humans while the baby is held by the surrogate mother. This is a relatively new process, and the results cannot yet be evaluated. Will the young, when reaching sexual maturity, desire mates from the species of the surrogate mother or from its own? Are cultural differences between species sufficient to inhibit social growth of adopted orphans?

These examples and questions tend to illustrate the extreme importance of the growing-up period, a time of change which prepares the infant for the rigors of life or, failing that, eliminates the poorer specimens.

Red fox pup playing

Life Styles

Man is a social animal. This overworked expression sometimes leads us to think that man is the only truly socially oriented species. Chauvinism, in this regard, is unwarranted. All higher classes of animals have species that are truly social or at least have species that exhibit some behavior that can only be interpreted as social.

What factor or set of circumstances has caused social adaptation to have such widespread occurrence? There are obviously advantages to social groupings and family living. Probably the most important has to do with predation. No animal species is totally free from predators, although some have defenses that make them practically invulnerable.

Predators can be better foiled by a group than by an individual because there are then present many sets of eyes and ears to detect a potential attack. The social group may also reduce overall stress in some species by allowing sentries to give other group members an opportunity to sleep, ruminate, or interact with group members while lowering their individual defenses. The practically motionless, ever-watchful Topi antelope sentry of the East African plains allows other group members to be less watchful.

Communal nesting is a characteristic of many bird species. While some parents are out searching for food, others stay back to help deter predators. Colony nesters also have fairly synchronous hatching and rearing schedules. The resulting abundance of young serves to gorge the local predators. Overall, fewer young are taken than would be the case if each bird nested individually and the eggs hatched at different times, producing isolated chicks that offered predators a singular target. Many ungulate species drop their young within a period of several days. Wildebeest literally flood the plains with their offspring so predators will remain well fed for critical periods, allowing a higher percentage of the young wildebeest to survive their most vulnerable period.

There are, of course, disadvantages to group living, but these have been effectively overcome by those species which depend on this structure. Competition for mates, food, nest sites, and nest material is increased. Parasites are more prevalent in large groupings. Some primates turn this to their advantage by forming larger groups during midday when biting flies are more of a problem. This assures that the number of bites per individual will be lessened. Infectious disease can also be more devastating in groups. However, this can aid a species' survival value by confining the disease and inhibiting its spread to other groups of conspecifics. It may also play some role in the establishment of antibodies to combat these diseases.

Close quarters may also make it more likely that other adult pairs, particularly birds, may kill the wandering young who enter their nest territory. Taking into consideration all these variables, however, it appears that social living may be an increasing phenomenon rather than a decreasing one.

Uganda kobs

The invertebrates are not unrepresented in social groupings. One of those we are most familiar with are corals. Corals often are the creators of atoll-type islands in warm-water areas. The laying on of layer after layer of living coral on the dead exoskeletons of past generations is essential in creating habitats for many species. The countless species of reef fishes, many of which are also colonial, depend on the habitat produced by these corals. And corals may eventually form islands that become hosts to colony nesting seabirds and even man.

Certain species of hydroids are also colonial, and the interrelationship between individual members is so great that scientists have given them the anthropomorphic name of "persons." Such "persons" form the feared Portuguese man-of-war "jellyfish," which is such a unified force that many think of it as a single organism.

Insects are also represented by colonial social groupings, the most illustrative of which are some bees, wasps, ants, and termites.

Many species of bees are definitely social insects. A single beehive may contain as many as eighty thousand individuals. The colony is made up of one adult queen, workers, and drones, as well as the larval stages which produce the recruits to replace lost members.

Larvae which are to become queen bees are fed a different quality and quantity of food by the workers. Any larva has the potential of becoming the queen, but the workers choose only one until the hive is ready to split and other queens are needed.

The workers tend to the cleanup, feeding, and wax production in the hive. They are the only ones capable of these chores. When workers return after locating a good source of food, they do a "dance" in a sort of figure eight which directs other workers to the location. This is a most important function of the social grouping, as it saves time for the individual workers.

The drones are the only males in the group, and their sole purpose is to fertilize the queen. They do not have a stinger because of their limited role in the life of the hive or colony. As a matter of fact, as the end of summer approaches, these now parasitic members of the social group are stung by workers and rejected from the hive.

The hive is a most complex structure and proper temperature and humidity must be maintained to protect the egg-laying queen. Bees regulate the temperature within the hive by spreading out when it is warm and piling on top of one another when it is cold. In extremely warm weather, the workers will position themselves at the entrance of the hive and fan with their wings to provide circulation and lower the temperature inside the colonial structure.

Termites are also well-documented social insects. The African savannah is marked by large termitaria. These structures provide food for the aardwolf and other insectivorous animals. They also provide cool daytime resting places for the aardwolf, the familiar wart hog, and a variety of reptiles. The aboveground portion of the termite mound may be two to five meters high and ten to twenty meters in diameter.

The division of labor in the termite colony is stratified, as is

Opossum mother with four young

that of the honeybee, but the termite is a much more primitive insect. It is an older species than the bee and has been a social animal for a much longer time. The queen of the termite colony shares her area with the king, and they are continually fed saliva by workers. The queen may produce tens of thousands of eggs daily. The immature members of the colony are divided into larvae and nymphs. The nymphs are the more mature stage of the larvae and resemble miniature adults. The nymphs grow by a series of molts in which the old exoskeleton is sloughed off and a new one formed.

Workers and soldiers make up the remainder of the colony. The soldier termites are larger and have more powerful jaws. Their task is to protect the colony. The workers are smaller and nonreproductive and assume the role of laborers, as do the workers in the beehive. Primary responsibilities of the termite worker are to maintain the temperature in the colony and care for the fungus gardens which provide food. The workers control the temperature to a practically constant thirty degrees centigrade. The termitarium contains many ribs or air channels that provide oxygen-rich air to the colony. The temperature is controlled by opening and closing these channels.

These unique groups of insects have formed very sophisticated colonial structures as their method of survival. A single termite or bee removed from its colony would be totally helpless and would quickly die. This is true, of course, only of the social species. There are solitary species as well.

Reptiles seem to be relatively free of social groupings. No examples of true social organization have been identified for reptiles. Some do tend to exist within what might appear to be colonies. The unusual marine iguanas tend to form such colonies on their home range in the Galapagos Islands. This may be, in part, to insure the availability of prospective mates. It may also be because the animals must be near beds of the aquatic vegetation that provide food. This, in fact, seems the more plausible reason.

Lioness and cubs

Spotted hyena pups

Hippopotami

Masai giraffe

Some snakes share winter denning sites. These hibernacula may even be shared by more than one species. This behavior may be more in response to a lack of suitable hibernation sites than because of any true social need.

Group living can have beneficial effects upon the evolution of a species. Evolution allows for "innovators." These are individuals who, by genetic accident or propensity for adaptation, can bring about behavior which may be passed on to other members of the species. In solitary individuals, the only way this can be passed on is through breeding or during the relatively brief period of parental care—brief, that is, for most species.

Because of the increased interest in the science of ethology— the evolution of behavior—there are many more field workers who are providing us with new information. An excellent example of this new awareness concerns a group of Japanese macaque monkeys which has been studied for over thirty years. The heroine of this group is a female who was named Imo by her human observers. In 1953, as an eighteen-month-old juvenile, she became the "innovator" of her troop.

The free-ranging group receives supplemental feeding from the researchers. This enables the scientists to control the troop's movements to some degree. An early food supplement was sweet potatoes. The macaques were fed in a sandy area near water. Imo, most likely by accident, dropped her potato into the water. In the process of picking it up, she washed off some of the sand. Within a month another juvenile was mimicking Imo's behavior and within four months her mother followed suit. Ten years later, approximately five of the group were using this behavior. The only individuals who weren't were some who already were adults at the time of the initial behavior discovery by Imo.

Behaviorists later threw peanuts into the water for the macaques to retrieve. Many members learned to swim and even to dive for food. This manipulation of behavior and troop movement by scientists is not an ideal method of observation and must be applied with caution. Researchers also began putting wheat out on the sand which some members of the troop would wash. Imo again became the innovator and was the first to throw sand and wheat into the water where the wheat would float and be easily harvested while the heavier sand sank to the bottom.

These facts do not mean that we are dealing with a "super" group of primates who may be evolving more quickly than any others. It simply means that this group has been sufficiently studied so that when a new behavioral pattern did occur, it was identified.

If Imo were a member of a solitary species, her newly acquired behavior would be severely limited, if not entirely lost. Thus the social group or community allows innovative behavior to be tested and rejected or retained with much more facility than in other types of societies.

The Adélie penguin, as far as sheer numbers are concerned, is certainly an important social species. A single breeding colony or rookery in the Antarctic may number from many thousands to over one million, and there are many such rookeries.

The male penguins are the first to arrive on the ice-locked beaches and they immediately squabble over territory. This bickering will go on until the nesting period has ended. Males begin their displays as soon as they arrive. First-time breeders try to attract females, and experienced males reunite with their old mates. When a pair has formed, the two birds stand face to face bobbing and weaving as an early strengthening of the pair bond.

Elephant defending young

Nest-building begins soon after pair formation, and pebbles are the only available material. One by one, these small rocks are carried to the nest site and heralded with mutual calling and display. The limited number of nesting stones encourages pilfering. Some individual stones are carried to several nests before coming to rest in a functioning nest with eggs.

Penguin pairs must pass other nests to reach their own in the crowded rookery and there is a great deal of jabbing and pecking

Herd of Burchell's zebra

Jackass penguins

along the way. There are a number of important behavior patterns and postures that enable these colony nesters to fulfill their role under such sardinelike conditions. Two eggs are laid, the second slightly smaller than the first and appearing several days later. After the second egg is laid, the male takes over incubation duty while the female, who has fasted for three weeks, heads to sea to feed. The way is not easy and leopard seals, as voracious as the African predator from whom they get their name, are waiting. Snow may completely cover the incubating males, but they do not give up their vigil and may go six weeks without eating. The females, after being at sea for two weeks, return to relieve their mates. During this period, the male's weight may be reduced by as much as one-third. He may stay at sea for two weeks feeding on krill, the small shrimplike creatures that make up his diet, until his lost weight is regained.

After five weeks of incubation, downy young Adélies hatch. One parent must always stay behind to protect the youngster from predator skua gulls (not true gulls). This dependency lasts for about three weeks until a crèche, or nursery, is formed. The chicks of the year huddle together for warmth and protection from predators under the supervision of a few adults who stay behind while most of the parents go to sea to seek food for their young. The parents feed only their own young, which they identify by voice. Even though it is so full it can hardly move, the chick literally chases its parents back to the water to gather more food. Those chicks whose parents do not return are easy prey for predators. As many as eight out of ten may not survive long enough to produce young themselves.

In seven to eight weeks the young Adélies molt, or lose their natal down, and grow the adult plumage they must have before they head to sea. Relatively few survive, but far more do than if they had not had the protection of the colony.

Social groupings, as in penguins, may appear to us to be more chaos than order, and the rapidly changing family groups of chimpanzees may not seem to have much of a pattern, but these time-tested methods are well refined. There is a price to be paid for social order, but the benefits are overwhelming. Only through intricate social structures can essential information be passed on to future generations.

Communal crèches, chimpanzee nurseries, and African lion play groups all allow interactions which pass along a well-defined body of information. This is only possible in social species. If we had evolved as a solitary species, we would not have the cities or other cultural structures that have arisen around the world. Perhaps the social group is a method that is still being tested. That is the function of evolution—constant refinement and adaptation.

Mallard hen and chicks

On Their Own

For most animals, the period of growth is far shorter than their tenure as adults. Many species of amphibians, insects, fishes, reptiles, and even one group of birds allow their offspring to fend for themselves. These are species that are on their own from the time of birth or hatching. Other groups, like the wolf spider, provide only minimal care before the young head off on their own. At the top of the scale animals like apes, cats, and elephants have long, arduous developmental periods to prepare their young for life.

But the final task is the same for all—the offspring must be equipped to survive to breed and carry their selected genetic material on into the future. So whether independence comes on day one or six to eight years later, the end result is the same.

Being on one's own, in nature, has many implications. Food is probably the most immediate need. Competition is the test. Food supplies are nearly always limited and have a definite inhibitory effect on animal populations. The bison, standing in a sea of grass, might appear to have no worries about food. But man's fences and space competition from conspecifics and other related mammals are factors. And there are years of drought when the green bounty is scarce. Snow may cover dry grass stems for months, and feeding can become a laborious pawing and digging with only meager rewards. Nature provides sustenance but rarely in unlimited quantity.

Shelter can be another pressing need. The young chimpanzee, who huddled at its mother's breast and was protected from cold and driving rain by her broad back, must one day seek its own shelter and tolerate the sting of the rain on its own back. A young elephant, when it is too large to slip between its mother's legs or even be protected by her shadow, must learn to protect itself from the sun. It must be a shock for the young kangaroo when it can no longer seek the warmth of the female's pouch.

Territory becomes a pressing need as the breeding season approaches. When the breeding urge is felt, the animal must try to respond. A home territory must be established and in some species a nest site or burrow system constructed. The young animal must, in many instances, rebuff rivals in order to pass on its genes. It may have to accept defeat in breeding competition in order to survive until the next breeding season when it may be stronger, more assured, and more attractive to females.

The male songbird may find himself driven out of his home territory by his own father. Young lions and bull elephants are forced from their families to live solitary lives or as members of all-male groupings until they are strong enough to challenge the leader for the right to breed.

All of these pressures are traumatic for the newly mature animals, and many young individuals, through ineptitude or lack of food or plain bad luck, do not survive. How many sea turtles even make it to the ocean? Predators must be especially well prepared for the important task of killing prey. Prey that is too large may kill a young hunter who has misjudged his capabilities.

Mortality rates for many species are virtually unknown. For insects, amphibians, snakes, and many other forms, we have only educated guesses. Large mammals, of course, are more readily studied, and the figures we have are more accurate. The extensively studied dall mountain sheep has an average life of nine and one-half years, which is approximately two-thirds of its maximum life span.

Rocky Mountain goat

Birds have been extensively studied, and although the figures we use may be speculative, they are almost certainly nearly correct. It has been estimated that for every hundred song sparrow eggs laid only seventy-five will hatch. Of these only two-thirds will reach the fledgling stage. And by the first breeding season, the survivors of the one hundred will be approximately six.

Open ground-nesting birds experience about a 60 percent hatch rate and 46 percent reach the fledgling stage. Hole-nesting species, because of better concealment and longer parental care, may hatch 77 percent of their young and fledge 66 percent. These figures can represent serious problems for birds that lay few eggs.

The reasons for this high mortality rate can be manifold. Inclement weather may inflict hardship on whole populations, and one study indicates that one species of rail, a small marsh bird, lost 50 percent more young in a year of severe storms than in a more nearly normal year.

Predators can also cause havoc in colony nesting areas. A single red fox on a breeding island of herring gulls caused a 94 percent increase in mortality. In another situation, a lone raccoon destroyed every nest in a tern colony consisting of five hundred mated pairs. Parasites and disease also play an important role in removing weak individuals from a population. Although the figures seem harsh, these high mortality rates are necessary to keep a species vigorous and inhibit overpopulation.

In one of the relatively few studies of amphibian mortality rates, —that done on the Great Plains toad—it was found that of the eggs laid only about one-third hatched. This, of course, does not compare favorably with hatch rates nearing 100 percent which are possible under laboratory conditions. Total survival in the species studied was thought to be about 10 percent.

The North American mountain lion, known by many names— puma, cougar, painter, panther, and catamount—is becoming rare over most of its range in the United States. The females generally produce young only every two years, and the dependence of the young upon the female is protracted.

Puma kittens are born spotted with ringed tails and black eartips. These markings eventually fade to a tawny color with only dark ear and tail tips remaining. Some South American pumas may be reddish or steel gray, and melanistic individuals do occur, if rarely.

Young mountain lions stay with their mother for up to two years and, as they mature, hunt with her to learn their skills. Even before they are fully grown, they may successfully take ground squirrels and other small rodents on their own. But deer, which will eventually be their primary food source, require more skill in

Baby kudu

the taking than can be learned in the first year and a half.

At two, young mountain lions may retain some shadow spotting, but as they approach maturity, at about three, the markings disappear. The female also becomes less and less willing to share her prey with them. Fierce scraps may occur and the young puma responds to an innate calling to be on its own. Pumas live a practically solitary life.

When the bond is broken, the most severe test occurs. Can the youngster kill on its own? Is the food supply abundant enough to feed the new hunter? Solitary hunters are more wasteful than predators living in a group or family unit. Although the puma is territorial, the male territories are not strongly defended. The males will scentmark and claw trees but generally avoid interlopers. At about three, the male will seek out a mate. The new pair will spend a couple of weeks together. They will hunt as a pair, sleep in contact with each other and mate repeatedly during the female's heat cycle. But once she has conceived, they go their separate ways.

The polar bear is an Arctic wanderer whose total world population may number fewer than ten thousand. The adult polar bear is almost totally marine, and only when a female has very young cubs does she avoid the icy sea water. Polar bears are born in dens under the snow. They weigh less than a pound at birth. The maturation process in polars is a long one, and although the cubs may stay with their mother through three winter dennings, two is more likely.

After it reaches two or three years the polar bear strikes out on its own, and it leads an almost solitary life once it parts from a surviving littermate. Although males and females must come together for mating, the rest of their lives are mostly spent alone; at least that is so for males. Breeding females generally have young with them in a family group.

Polar bears spend most of their lives seeking food—seals, lemmings, and fish.

The newly independent polar bear is most likely to come in

Yellow-shafted flickers

Bobcat

Pronghorn

Cottontail rabbit

contact with others of its kind at the site of a beached walrus, seal, or whale carcass. A whale offers an enormous feeding opportunity, and many polar bears may feed and rest close to the carcass for several days; still, there is no polar bear society. Newly matured polar bears learn to den up when heavy storms cause an arctic "white-out" that makes travel impossible. They also learn to move from ice floe to ice floe, for it is on the ice, not on land, that the major food supply is to be found.

The Delmarva fox squirrel is listed by the United States Department of the Interior as an endangered species. It is considerably larger than the gray squirrel which often shares its habitat. The gray squirrel prefers forest with dense underbrush, whereas the fox squirrel, which spends much of its time on the ground, prefers open cover under large trees, such as the loblolly pine. Where fox squirrels and gray squirrels appear in the same territory, they can coexist by utilizing different levels of the habitat. The problem for the fox squirrel occurs when it is forced to accept marginal habitat.

When young male fox squirrels reach maturity, they are forced out of the parents' territory and must fend for themselves. This almost invariably forces them into a habitat not ideal for their survival. The limiting factor for the Delmarva fox squirrel, and most other endangered species, is lack of habitat.

The young male fox squirrel generally stakes a claim on the periphery of desirable habitat. To expand his territory toward proper feeding and foraging areas would impinge on other established fox squirrel territories. This pattern is true unless a young animal is extremely fortunate and can take over a territory after the previous owner has died or been removed.

Delmarva fox squirrels are fond of man's agricultural products and raid his corn and soybean fields. Of course, because of their marginal territories, they may venture farther and farther from protected cover to seek food. This exposes them to predators, particularly hawks, who take them when they are in the open. The only young of the year likely to survive are those that can take over an abandoned territory.

The independence of a young animal is earned by a long and complicated set of circumstances. Skill, ability to learn, and even fate have played important roles in getting the newly adult animal to this point.

The new adult represents the best that nature has to offer for that given territory, for that given season, and for the species. The giant Kodiak bear, the world's largest living land carnivore, is found only on the Kodiak and Afognak islands off Alaska. The number of cubs produced on the islands every year is limited by territory and by the amount of food the islands can provide.

When the young adult animal, no matter what its species, starts out on its own, it is by then among the strongest and most adaptable of its kind. Nature has provided it with tools which have been sharpened by natural selection. Each day, as the young animal approaches adulthood, its chances for survival improve. Its increasing skills make it one of a select group of individuals who must carry on their unique genetic potential. This one animal has bested the challenge of the growing-up period.

If there is a reason, a "pay-off," an essence to biological existence, it occurs as the young, but now-adult animal, approaches the task of creating the next generation. First it was born. Then it grew and learned and survived. Now, with luck, it will stamp the future with the genes that have proved themselves to be sufficient to the needs of its species. Survival of the fittest does appear to be the way of the natural world.

Leopard

Photographer Credits

Design by Louise Crandell and
Nigel Rollings.

Color reproductions produced at
Studio Analysis under the personal
supervision of Paolo Riposio,
Torino, Italy. Represented in the
United States by Offset Separations
Corporation, New York.

Set in Palatino by American–Strat-
ford Graphic Services, Inc., Brattle-
boro, Vermont.

Printed and bound by Kingsport
Press, Kingsport, Tennessee.

An AG Edition